RIDING THE SHEEP'S BACK

RIDING THE SHEEP'S BACK

The Rise and Fall
of John McNamara and Company

John Brenan

with associates

Andrew Brenan & Helen Birch

The Publisher's Apprentice

PO Box 224W
Ballarat VIC 3350
sales@connorcourt.com
www.connorcourt.com

ISBN: 9781922168740 (pbk.)

Cover design by The Publisher's Apprentice

Printed in Australia

CONTENTS

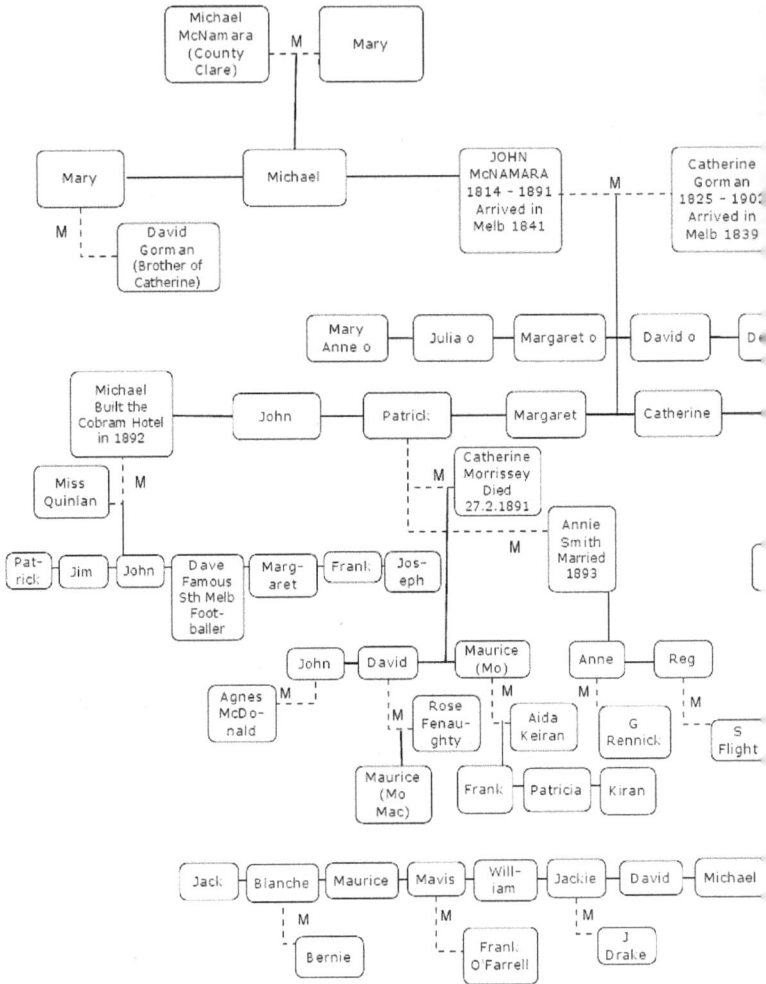

Key

- - - M = Married

_____ = offspring
& siblings

o = Died in Infancy

Catherine
Gorman
25 - 1902
rrived in
Melb 1839

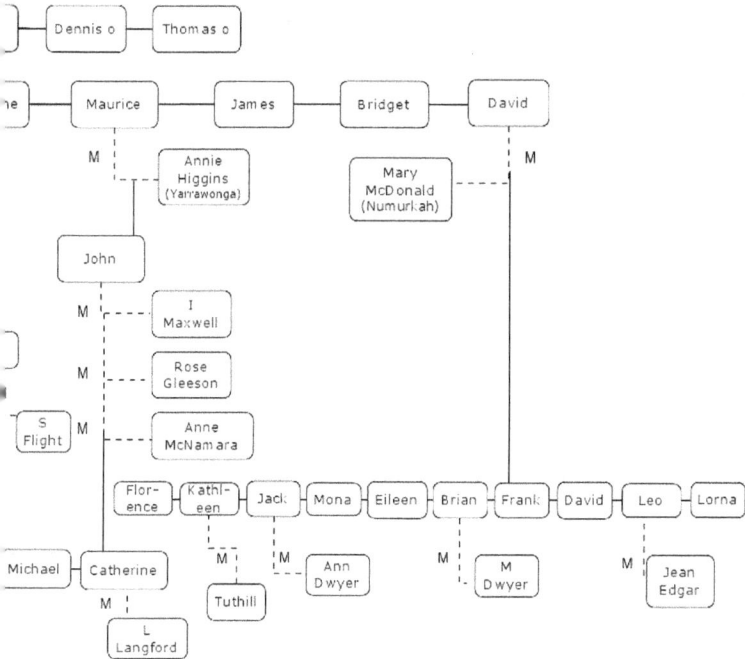

Dennis o — Thomas o

he — Maurice — James — Bridget — David

M

Annie
Higgins
(Yarrawonga)

Mary
McDonald
(Numurkah)

M

John

M — I
Maxwell

M — Rose
Gleeson

S
Flight — M — Anne
McNamara

Flor-
ence — Kathl-
een — Jack — Mona — Eileen — Brian — Frank — David — Leo — Lorna

Michael — Catherine

M — M — Ann
Dwyer

M — M
Dwyer

M — Jean
Edgar

Tuthill

M

L
Langford

This book is dedicated to John and Anne McNamara and to their daughter Marianne who has all their character and charm.

Introduction

This book is not a genealogy of the McNamara family, but is a history of the iconic family company of stock and station agents that was well known in the early 20th century.

Throughout the book, many names, such as John and Maurice, are repeated and confusing. To help with this, we have included a family tree, which was prepared by the late Jim McNamara, with the help of Emma Nairn. Jim's original work has been modified and abridged to make it more relevant to the history of McNamara and Company.

Unfortunately, many of the original company records have been either lost or destroyed over the years. This includes all records of McNamara directors' meetings.

Brian Rodwell, who managed McNamara's for Australian Estates after the sale of the company, does not recall seeing any minutes or company records. Lyn, of the Marketing department of Elders, also carried out an extensive search of their archives and found nothing.

The company solicitors, Maurice Blackburn and Company, destroyed their records after seven years, as was the usual practice. In the days before computers, no digital records or other backup of information was kept.

The many sources used in compiling this history are listed.

However I particularly wish to mention Pat O'Connell, who provided wonderful insight into the latter days of the company, and Mark McNamara, who provided the fascinating memoirs of the early life of Maurice McNamara (Mo).

I also wish to acknowledge the contributions of my associates: Andrew Brenan, who organised all the illustrations and documents, and Helen Birch, who did much of the early research work into shipping and passenger lists. It was Helen who, thinking laterally, discovered the marriage certificate of John McNamara and Catherine Gorman, where John was only listed as "Mack". She also provided great support in the production of this book.

PART ONE

A Snapshot of Irish History

1

A snapshot of Irish history and its impact on Australia, particularly Victoria. This helps in discussing the history of John McNamara and Company.

During the reign of the first Queen Elizabeth, between 1558 and 1603, at least three wars were fought in the Irish Province of Munster.

The First Desmond Rebellion (1569-1573), the Second Desmond Rebellion (1579-1583) and the Nine Years' War (1594-1603) eventually resulted in the defeat of the Irish by Lord Mountjoy, and Catholic Ireland being under the control of Protestant authorities.

The estates of the Irish Earls, who fled to Europe in 1607, were forfeited to the Crown, which helped to further consolidate Protestant power in Ireland.

In 1649, Oliver Cromwell arrived in Ireland and devastated the country to avenge the killing of Protestants in the Irish Uprising of 1641.

The hostility that had originally started between Catholic Ireland and Protestant England continued between Catholics and Protestants in Ireland.

This hostility was only aggravated by the later Orange invasion and battles such as the Battle of Boyne.

The Act of Union

The bloody revolution of 1798 of United Irishmen, led by Wolfe Tone, and the uncertainty that followed, caused the English government to introduce The Act of Union in 1800. This dissolved the Irish parliament, so that all government then resided in Westminster.

The introduction of the Act of Union only created more disruption and was another factor driving rebellions calling for Home Rule and later for a separate Republic.

By 1843 Daniel O'Connell sought a repeal of The Act of Union and was able to mobilise the Irish people by mass meetings. These meetings were organised through local pubs and became the basic contact with peasants, who paid "the Catholic rent" of one penny monthly.

Transportation of convicts to Australia

The first contact between Ireland and Australia was when the First Fleet of convicts, and the soldiers who guarded them, arrived in the new country. Transportation of convicts from Ireland had the dual purpose of providing a labour force in Australia and at the same time punishing felons.

In total there were 40,000 to 50,000 Irish convicts, a quarter of whom were women. Most convicts were felons charged with crimes of cattle stealing or destruction of landlords' property, crimes which were partly politically motivated. There were also those who had been convicted for purely political reasons following rebellions.

1,800 convicts were transported to New South Wales as a result of the 1798 bloody rebellion by United Irishmen.

Years later, the failed 1848 revolt by Young Ireland sent leaders, including John Mitchel, William Smith O'Brien and Thomas Francis Meagher, to Van Diemen's Land (Tasmania). Mitchel and Meagher later escaped and emigrated in the early 1850s to the US, where Meagher served as a Confederate General in the American Civil War.

Passage from Ireland

In 1820, the penal laws that had been established in 1695 by William of Orange, following the battle of the Boyne, were repealed, and so Catholic emancipation no longer influenced migration.

Two-thirds of the Irish coming to Australia in 1830 were convicts and the other third were free settlers. By 1850 this ratio was reversed and two-thirds of Irish immigrants were free settlers. Most of these came from Munster, Clare, Tipperary, Limerick and Kilkenny, which is in Leinster.

The average migrants were young male, female or family groups from rural backgrounds, either unskilled labourers or domestic servants. As crop farming was less profitable, farm labourers' experience was mostly gained through animal husbandry.

A large proportion of migrants also went from Ireland to North America, paying £2 for passage in very old vessels, later described as "coffin ships".

Australia was a more expensive destination, requiring subsidising through a Bounty Scheme by British or Australian Colonies.

In the 1840s, the Irish coming to Victoria tended to be Protestants. Later, in the 1850s, the majority were bounty or subsidised migrants.

Those migrants arriving before the Irish famine tended to be large family groups seeking to increase their opportunities. Females of marriageable age were not restricted in Australia by the 'dowry' system, as many were in Ireland, and so had greater opportunity for marriage.

When they did marry, the Irish tended to have large families. This was a result of using no contraception, but also stemmed from needing large numbers for support in their old age.

When Irish migrants came to Australia, they continued to have large families, commonly with rates of 10-15 births per family, although death rates among children were also high.

Expansion of the Irish population and pressure to emigrate

From the time of the Act of Union in 1800 to the time of the Great Famine, which began in 1845, the Irish population increased rapidly. At the time of the 1844 census, the population in Ireland had reached over eight million.

This population expansion increased the pressure to emigrate. However, a decision to emigrate to Australia would mean that the migrant faced a 3-4 month sea passage in difficult and often intolerable conditions. Ships were privately owned and used both for convict transportation and emigration, although on different voyages.

Land ownership problems in Ireland

The hostility between Catholics and Protestants that resulted from the earlier wars was an important matter affecting Irish tenants. Many parcels of land owned by absentee Protestant landlords had been subdivided into smaller and smaller lots, and the high rents were collected by agents on behalf of the landlords.

Realising the problems with this arrangement, the English government set up the Devon Commission in 1853 which reported:

> It would be impossible to adequately describe the privations which the Irish labourer and his family habitually and silently endure. In many districts their only food is the potato and their only beverage is water. Their cabins seldom give protection against the weather, a bed or a blanket a rare luxury, in nearly all a pig and a manure heap constitute the only property they have. Suffering greater we believe than any other people of any other country in Europe.

The average Irish family had no possibility of owning their own land.

Emigration during the famine years

Between 1845 and 1852 the potato blight fungus devastated Ireland's population, bringing one million deaths and causing emigration from Ireland of around two million – mainly to Britain and America.

During the famine many landlords evicted their poor starving tenants who failed or were unable to pay the rent. By 1851 over 200,000 starving families lived in overcrowded poorhouses or workhouses. Those weakened by starvation succumbed to fever, including orphan children.

Australia received about 20,000 migrants during the famine years. In the decades following the famine, 300,000 Irish came to Australia, these migrants mainly prompted by a stagnant Irish economy.

A memorial to the Irish famine victims was later erected in Williamstown.

Intolerable conditions on board

For many, conditions on board the ships that carried them from Ireland to Australia were intolerable.

Passengers were afraid to take the recommended daily exercise above decks, which compounded the claustrophobic conditions below decks. This, along with the necessity to batten down hatches and remain below deck during days of relentless storms, was the cause of many health problems and deaths on the long voyages.

A government enquiry into conditions on the ships reported:

> That in many cases of passengers not going on deck their health suffered so much so that their strength was gone and they had not the power to help themselves. The 'tween decks was like a loathsome dungeon and the stench was like from a pen of pigs. Deaths were usually of young children deprived of proper diet.

On the larger ships, a surgeon superintendant was in charge of the health and welfare of passengers. He received ten shillings for every landed passenger.

The Irish in Australia

For those Irish migrants coming to Australia – including John McNamara and his future bride, Catherine Gorman – the scene on arrival must have been shocking.

Melbourne had only been designated "as a place for a village" in 1835 and migrant accommodation was provided by tents, in Canvas Town in South Melbourne or Tent City in Bourke Street.

In Melbourne they settled around Lonsdale Street, west of Elizabeth Street. Protestant migrants, including doctors, lawyers and teachers, chose to settle in Hawthorn, Kew, Prahran and St Kilda.

The "tent cities" of Canvas Town and Tent City became unhealthy slums and the priority was to open up Crown Land to selectors.

In other parts of Victoria, Irish migrants tended to cluster with migrants from the same counties in areas around Kilmore or in the west around Warrnambool, Koroit and Belfast (Port Fairy).

The population of Victoria jumped from 10,000 in 1840 to 500,000 in 1860. Melbourne, which had begun as a Tent City in 1851 with a population of 29,000 (17% of which were Irish immigrants) increased its population to 123,000 by 1854.

On 1 July 1851 Victoria became a separate State. Days later, James Esmond discovered gold in Clunes. Gold fever then became a large factor affecting the rapid rise in population.

Land selection in Australia in the early years

A Land League was established in 1856, developing a policy of selection before survey of 160-230 acres for instalment repayments. Influenced by squatters in 1857 the Legislative Council rejected the Act. Later, in 1857, the Selection Act (selection after survey of 320 acres at £1 per acre on time payment) was accepted.

In 1859-60 a modification took place, with land divided into two parts, half purchased at £1 per acre and the other half at one shilling per acre annually for 7 years with the right to purchase.

Selectors were prevented from immediate resales of land. They were required to be resident on lands and to carry out improvements. Wooded blocks had to be cleared and sown to pasture.

Squatters continued to oppose selection.

Many land applicants ensured their relatives selected adjacent blocks to consolidate property.

Charles Gavan Duffy

The arrest of Charles Gavan Duffy, who went on to become Premier of Victoria, is a noteworthy event for Australians. As a leader of Young Ireland, he had been unsuccessfully tried for treason five times before all charges were dropped in 1849.

Duffy entered the English parliament in 1852 where he worked for Irish land reforms and compensation for those tenants who improved the land. In failing health he moved to Victoria in 1856.

Duffy was elected to the Victorian Parliament and appointed Minister of Lands (1858-1859 and 1861-1863). He used his Irish experiences to organise the pressing need for distribution of Crown Lands, favouring the small selectors over the large land holding squatters.

Charles Gavan Duffy became the Premier of Victoria in 1871 and was knighted by Queen Victoria in 1873.

Irish 'baggage'

From their Irish homeland the migrants brought along some additional 'baggage' to Australia.

Firstly, their habit of rebellion flowed over to the new country. In 1804 a small rebellion of convicts at Castle Hill in Sydney was reported, and later, in 1854, Irish immigrant Peter Lalor led the Eureka Stockade.

In the 1860s the Fenian scare followed the Fenian revolt in Ireland in 1866 and subsequent attempted assassination of the visiting Duke of Edinburgh at Clontarf (1868). Later in the 1880s, the Kelly outbreak is said by many to have had its roots in the Irish heritage of Ned Kelly's supporters and others involved.

More importantly, there was the antipathy and hatred between Catholics and Orangemen in Ireland that spread like a virus to Australia. Until relatively recent times, anti-Catholic and anti-Irish bigotry was rife here.

The Irish convicts were regarded as *untermenschen*. In Samuel Marsden's history of the Irish in Australia, he describes them as "an unaccountable set of beings".

In the 1840s and 1860s there were riots between Catholics and Protestants in Melbourne. Early elections for the Melbourne City Council, in 1841, were controlled by Freemasons and Orangemen.

Accentuating the intolerance in the 1880s was the anti-Catholicism of the Orange Lodges in Ulster.

The final Irish revolt: the Fenians

The final Irish revolt in the 19th century occurred in 1866. This was a revolt led by the Fenians, who had infiltrated the Irish-based army and hoped for help from Irish Fenians who had fought in the American Civil War.

This revolt collapsed and its leaders were transported to Fremantle, Western Australia, on the convict ship *Hougomont* in 1868.

Several years later, in 1876, using a whaling vessel the *Catalpa*, American Fenians arranged and executed a most daring rescue of six of the Fenian convicts.

The "Hougomont" was the last convict ship to come to Australia – incredibly only 32 years before Federation.

General references used throughout this chapter

Marie & Conor Cruise O'Brien, *Concise History of Ireland*, Thames and Hudson, 1973.

Prof. Elizabeth Malcolm, University of Melbourne, Courses 131219 and 30048: "Modern and Contemporary Ireland Since 1790"; "Ireland Down Under 2011".

Cecil Woodham-Smith, *The Great Hunger*, Hamish Hamilton, 1962.

Don Charlwood, *Settlers Under Sail*, Victoria Press, 1991.

Jeff Brownrigg *et al*, *Echoes of Irish Australia*, Galong, 2007.

Peter Stevens, *The Voyage of the Catalpa*, Carroll & Graf, 2002.

PART TWO

The Founders of
John McNamara & Co

2

In 1840 and 1841, Michael and Mary McNamara, farmers from County Clare in Munster, sent their three children, John, Michael and Mary to Australia.

At the time, Clare was the top Irish county sending migrants to Australia. McNamara was the second most numerous surname in County Clare, with the Master Index of Baptisms during that time recording just over 2460 McNamara families.[1]

Since the late 11th century, the McNamaras had been lords of Clan Cullen, a greater part of East Clare. There were also other septs – or divisions – of McNamara's found in other counties, particularly Limerick, Mayo, Tipperary and Cork.

When Michael and Mary McNamara's children left Clare for Australia, they were allocated travel on different ships. The younger children, Michael and Mary, were £18 bounty migrants. They sailed on the *Neptune*, which left Plymouth on 12 December 1840, arriving in Melbourne on 27 March 1841.

Their older brother John, aged 27 years, a farm labourer and carpenter, migrated on a £19 bounty. He sailed on the *Duchess of Northumberland*, a barque of 550 tons departing Plymouth on 17 February 1841, and arriving in Port Phillip on 3 June 1841.

The voyage of the *Duchess of Northumberland* that brought John to Australia was an historic trip as it was the first ship to bring mail direct from England to Port Phillip.

As with other ships, the *Duchess of Northumberland* provided charter as a convict ship on one voyage and migrant ship on another, sailing as a convict ship to Hobart in 1842 and 1853.

John and Catherine McNamara (nee Gorman)

A few years before the McNamaras emigrated from Ireland, the Gorman family, from Mitchellstown in Cork, just over the border from Limerick, had also arrived in Australia. They travelled on the *Wiliam Metcalfe*, which left Plymouth on 26 July arriving in Port Phillip on 15 November 1839.

One of the Gorman daughters was 15-year-old dairymaid Catherine (1824-1902), who had travelled on an £18 bounty as the younger McNamaras had. Also arriving on the *William Metcalfe* was

WILLIAM METCALFE (ALSO KNOWN AS WILLIAM METCALFE)
1832 - 1845
Reproduced by permission of the Maritime Museum, Greenwich

DUCHESS OF NORTHUMBERLAND - Arrived Melbourne June 3rd 1941

No	Name	Age	Trade	Religion	Read and Write	Native Place	Amount Bounty	Engaged by	Employ	Rate of Wages	With or No Rations
52	MC NAMARA John	23y	Farm Servant	R/C	Both	Clare	£19	Nil	Nil	Nil	Nil

WILLIAM METCLAFE - Arrived Melbourne November 15th 1839 initiated by Mr.John Marshall London.

Name	Age	Trade	Religion	Read and Write	Native Place	Amount Bounty	Engaged by	Employ	Rate of Wages	With or No Rations
GORMAN James #	28y	Farm Serv.	R/C	Yes	Limerick	£18 per head	Capt Brown	Melbourne	£30 per annum	with rations
GORMAN Bridget	25y	Dairy Maid	R/C	-	Limerick	£36 family				
GORMAN Margaret	4y		R/C	-	Limerick	paid for				
GORMAN David	3y		R/C	-	Limerick	paid for				
GORMAN Patrick #	10m		R/C	-	Limerick	paid for				
GORMAN Judith #	10m		R/C	-	Limerick	paid for				
GORMAN Patrick	39y	Agriculturlist	R/C	-	Mitchelstown	£18 per head	Capt.Brown	Farm Lab	£75 per annum For Family with	
GORMAN Mary	35y	Dairy Maid	R/C	-	Mitchelstown	£18 per head	Capt.Brown	Dairy Maid		
GORMAN John	8y		R/C	-	Mitchelstown	paid for	Capt.Brown	Cow Boy		
GORMAN Mary	10y		R/C	-	Mitchelstown	paid for	Capt.Brown	Servant		
GORMAN Catherine	15y	Servant	R/C	-	Mitchelstown	£18	Capt.Brown	Dairy Maid		
GORMAN David	18y	Farm Serv.	R/C	Yes	Mitchelstown	£18	Capt.Brown	Labourer		
GORMAN Patrick # Unmarried Male	28y	Farm Serv.	R/C	Yes	Mitchelstown	£18	Capt.Brown	Farm Lab	£30 per annum	with rations

NOTE : 2 Sets Twins #   #

stockbroker J.B. Were, bringing with him his family, two servants, a tent, a fully prefabricated house and £9000 for investment.

On her arrival, Catherine lived in Melbourne, which had been founded only four years earlier. John McNamara would have lived in Tent City and must have been close to the Gormans. He and Catherine were married at St Francis' Church Melbourne on 18 January 1843.

Together John and Catherine had seventeen children, five born in Tent City who died in infancy. It is likely that John supported the family by working as a carpenter.

John's sister Mary (1832-1910) married Catherine's brother, David Gorman (1821-1901).

According to Kelly historian Ian Jones in his book *A Short Life*

Best Quality image
Damaged Page

Roman Catholic MARRIAGES celebrated in the District of Port Phillip, from the year 1842, to the year 1843, as recorded in the Registrar General's Office, at Sydney, New South Wales.

NO	NAME OF BRIDEGROOM	NAME OF BRIDE	WHEN MARRIED	WHERE MARRIED	NAMES OF WITNESSES	BY WHOM SOLEMNISED	No. in SYDNEY REGISTER
325	Dyer Bartholomew	Blaa Honora	21 November 1842	Melbourne		Rev. P. B. Geoghegan	
326	Honessey James	Lovell Anne	23	do		do	
327	Atkins James	Donohoe Catherine	15 December 1842	Geelong		Michael Stephens	
328	Corcoran Michael	Phelan Anne	26 June 1842	Melbourne		P. B. Geoghegan	
329	Cosgfor Patrick	Cooney Honora	24	do		do	
330	Magrath Edward	Delahunty Catherine	5 January 1843	do		Darrell hia? Grey	
331	Maidment Joseph	Sheenahan Margaret	8	do		do	
332	Smith John	Loukon Mary	8	do		do	
333	Sullivan Thomas	Rowan Anne	8	do		do	
334	Clarke Charles	Garey Margaret	8	do		do	
335	Mack John	Gorman Katherine	8	do		do	
336	Box William	Dillon Sabrina	8	1843	do	do	

I Certify, the foregoing to be true Extracts from the Original Register of *Roman Catholic* Marriages

John McNamara and Catherine Gorman's marriage certificate.
With John listed as 'Mack John' (2nd from bottom, left) these records were not easy to find

of *Ned Kelly*, "John Kelly had built a house on a half acre town allotment at Beveridge on the southern flank of Big Hill (Mt Fraser). It was here that Ned Kelly was born, December 1854, with the help of midwife Mrs David Gorman (nee Mary McNamara), who lived on the eastern flank of the hill.[2]

Eight or ten years after his marriage to Catherine, John was able to select 260 acres in the Chinton Parish of Springfield, an area between Kilmore and Lancefield in sight of the Macedon Ranges.

John Snr farmed in the area until 1869 when he selected land of 300 ¾ acres on the Goulburn River at Toolamba. He sold the Chinton Springfield land in 1874.

Catherine McNamara (née Gorman) at 'Springfield', Numurkah

The property at Toolamba was financed by a mortgage with the Executors and Agency Company of Melbourne, a company that was the first of its type, but which became insolvent in the 1980s.

Many years later, John and Catherine's son Patrick also selected 157 acres of heavily timbered land at Toolamba, and their son Michael selected an area known as "The Lowlands", which he later resold to his brother Patrick.

Approaching Toolamba. The McNamara property is to the left of the road sign. The treeline at the back marks the Goulburn River

Having retired from farming in Toolamba, John McNamara Snr died in Yarrawonga in 1891.

It was three of John and Catherine McNamara's sons, John [1855-1927], David [1863-1923] and Patrick [1845-1922], who were involved in building the John McNamara Company.

John McNamara Jr [1855-1927]

John McNamara Jr was born on the family's Springfield property, which was known as McNamara's farm.

After early life in Springfield, in the 1870s John Jnr moved to Gippsland and became involved in the horse trade to India around Maffra, Sale and Heyfield.[3]

Original Will of John showing how he succeeded in Australia.

PROBATE AFFIDAVIT.

"Although the said deceased is described herein as a Farmer he was not at the time of his death or for some years previously engaged farming, but was living privately.

Sworn at Yarrawonga in the Colony of Victoria this 29th day of May 1892.

Signed: *Catherine McNamara.*

In the Will of John McNamara formerly of Toolamba late of Yarrawonga the Colony of Victoria –

Assets: Real Estate.

---

1. Allotments 27 and 29 Parish of Toolamba
County of Rodney containing 300 ¾ acres.
Fences, post & rail, post & wire and chock & log.
Land subdivided into 4 paddocks 270 acres
broken up for cultivation.
Large tank, Weatherboard building containing
3 rooms, detached Kitchen of 2 rooms, Stables,
Yards and Outbuildings.   Value   £ 1804.10. 0

2. Part of Crown Allotment 18 Parish of Yarrawonga
containing 4 acres 3 roods 23 perches.
Four roomed Brick House containing, Parlor 11 feet
by 13 feet, 2 Bedrooms 12 feet by 10 feet, Dining
Room 11 feet by 13 feet and Passage through centre
4 feet 6 inches wide.   Walls of brick, floor soft wood,
ceilings wood, roof iron. Verandah in front and at side.
Weatherboard detached Kitchen 2 rooms, Iron roof.
1 stall stable and buggy house. 30 chains fencing
post and wire.   Value   £ 250. 0. 0

3. Part of Allotment 18 Parish of Yarrawonga
containing 8 acres 1 rood 28 ½ perches. 42 chains
of post & wire fencing.   Value   £ 100. 0. 0

Livestock –
Horse £5. Cow £ 3. 5.0. Calf £ 1   £   9. 5. 0
Carriages & Buggy   27. 0. 0
Harness & Saddlery   3. 0. 0
Furniture   24.15. 6

                                          £ 64.10. 6

            Total   £ 2218.10

John Jnr had a great reputation as a crack horseman and was able to quieten the wildest brumby. In the late 1870s he turned his attention to the stock agency business. As this was very competitive in Gippsland, he moved to Yarrawonga where stations were rapidly being subdivided into farms, taking a job as pound keeper.

In 1881 John approached the manager of the newly formed Australia New Zealand Bank and asked for money to start a stock agency. The bank manager apparently asked him how much he wanted and John answered, "How much have you got?" He started

The first McNamara in the stock agency business, John McNamara,
who founded his firm in Yarrawonga in 1881

his own business as a livestock and real estate agency and with £296 built an auction mart and office in Belmore Street, Yarrawonga.

The Yarrawonga business prospered and John's brother David came to help manage the firm. In 1887 John purchased 1792 acres in Coreen over the Murray River in New South Wales. This was the foundation of his well known Emu Park Estate.

Around this time, John's younger brother Maurice [1857-1885], a butcher in Yarrawonga, died of typhoid fever, aged just 28 years.

The original headquarters of the McNamara family stock agency business in Yarrawonga

Two years previously Maurice had married Annie Higgins [1865-1949]. Their son John Joseph McNamara [1885-1967] was born in the same year his father died.

Annie, aged 20 and possibly suffering from post-natal depression accentuated by the death of her husband, had her son adopted by his uncle John Jr, and so John Joseph was brought up at Emu Park.

At the age of 40, John Jnr had married Catherine Byrne in Wangaratta and together they had six children.

John Jnr retired from taking an active interest in the business in 1912, and remained at Emu Park until five months before his

2,000 one-year-old crossbred ewes make the first crossing over the new stock route at Yarrawonga weir, 19 November 1940

death, when he moved to his property "Mon Repos" in Park Street St Kilda. He died on 19 April 1927.

John Joseph McNamara [1885-1967]

Raised from infancy by his uncle John Jnr, John Joseph McNamara was groomed to take over the Yarrawonga branch.

In 1904, aged 19, John Joseph, together with others forced by drought, drove a mob of 6,000 sheep from Yarrawonga to Wilcannia and back of Bourke for feed. The drive continued down by the Darling River to Wentworth then back to Yarrawonga, the expedition taking six months to complete.

In Yarrawonga there was no adequate market for sheep so they were taken up to the mountains near Mansfield and let loose. However when the men returned for the sheep they could only round up 3,000, although fortunately their sale did cover all expenses for the whole drive.[4]

Edward McConville (left) and John McNamara (nephew of the founder)
who were in partnership for 33 years

John Joseph McNamara took over the Yarrawonga business initially as sole proprietor-auctioneer salesman in 1912. In 1919 he was joined by Edward McConville and their happy partnership continued for 33 years.

The Yarrawonga business prospered for many years with the increasing wealth in the area until in 1952, together with other McNamara businesses, it was sold to Australian Estates. At that time, one of John Joseph's sons, Michael (Mick) McNamara, took over management of Yarrawonga branch for Australian Estates.

In 1978 Mick purchased the original premises in which his great uncle started the business back from Australian Estates and started his own agency as McNamara and Co. That business ceased to exist upon Mick's death in 2007.

. M. J.

☒

Of Your Charity

the repose of the soul of

Capt. John Francis McNamara,
M.C.,
Killed in action in New Guinea
December 8th, 1943
Aged 34 years.
R.I.P.

"We have loved him in life, let us
not forget him in death."
—St. Ambrose.
"Eternal rest grant unto him, O
Lord. and let perpetual light shine
upon him."—300 days' indulgence.
"May he rest in peace."
E. J. DWYER, SYDNEY & BRISBANE

☒

Of Your Charity

Pray for the repose of the soul of

Maurice David McNamara
LATE A.I.F.
Died at Wangaratta
17th May, 1948
Aged 35 years
R.I.P.

"Jesus, meek and humble of heart,
make my heart like unto Thine."
(300 days' indulgence).

John Joseph McNamara was married three times. He had five children from his first marriage and a further five from his third marriage. One of the sons from the first marriage was John Francis McNamara [1909-1943].

John Francis McNamara [1909-1943] ("Big John")

John Francis McNamara was one of the most attractive and popular agents in the Yarrawonga and Wangaratta areas. He was known as "Big John", being six foot five inches in height.

Big John left the family firm in 1940 to join the Australian Imperial Force. In 1942, while serving in the Middle East, he was awarded the Military Cross for gallantry.

In 1943, while fighting the Japanese in New Guinea, Big John was killed in action (see Appendix 1). His brother Maurice also joined the AIF, serving as a Lieutenant and later dying of war wounds in 1948.

Michael McNamara [1843-1904]

Michael McNamara was the eldest son of John and Catherine. Born in Tent City in 1843, Michael selected land in Toolamba known as "The Lowlands" which he later sold to his brother Patrick.

He then farmed in Boosey, where in 1870 he met and married Mary Quinlan. They moved to Cobram where Michael built the Majestic Cobram Hotel in 1892.

Michael and Mary had nine children, including son David John McNamara [1887-1967], the famous St Kilda footballer who kicked a football world record of 93 yards (85m) in 1923.

David McNamara [1863-1923]

David McNamara was born in Springfield. Later in his life he named the family home in Numurkah "Springfield".

After his early life and schooling, at the age of 19, David joined his older brother John in Yarrawonga. Shortly after, a decision was made to open another office in Numurkah and in 1889 David took over this independent business.

David also opened offices in Nathalia and Cobram. The business prospered under his leadership together with his partner Tom Hurley.

It was said that David led something of a trend among

The Numurkah office started by David and John McNamara

auctioneers by cupping his hand to the ear to assist his hearing during the auctioning process.

At the age of 25, in 1888, David married Mary McDonald, and over the next twenty years they had ten children. Tragedy followed in 1907 after the birth of their last child and Mary became very depressed, tearful and undoubtedly suffered post natal depression.

Mary received psychiatric treatment in Melbourne during 1907 and was then admitted to the Bonleigh Private Psychiatric Hospital Elsternwick in February 1908.

Although not thought to be suicidal, Mary had to share a room with another woman and her baby. This seems to have triggered her suicide by hanging on 22 February 1908.

The Coroner's Report attributed Mary's symptoms to malnutrition. At that time, post-natal depression was not a recognised diagnosis and there was no real treatment available.[5]

David and Mary's eldest son John was to be a director of the family company.

Their second son, also David, was killed in 1917 while fighting with the First AIF in Flanders.

Leo (born August 1902) worked in the Numurkah office, with his father and Tom Hurley. Leo married Jean Edgar and had four children, including Margaret Burke. He started work at the age of nineteen and died at age 65.

Son Frank [1900-1976] worked as a businessman, possibly with the company. And David and Mary's son Brian was never associated with the company but ran a hotel in Kyabram.

In 1918 David Snr was married for a second time, to Grace Brennan. He died in 1923, in Irving Street, Malvern.

David McNamara and family after Mary had died

John McNamara, 1930

John McNamara [1892-1931]

John McNamara was the eldest son of David and Mary.

In 1916, at the age of 24, John was sent to Melbourne to manage the brothers' business. It is likely that his father David expected him to continue as Managing Director when the new company was set up in 1922. John did remain as a director, but was supplanted by Maurice McNamara ("Mo").

During the rest of his time there, with the company under Mo's management, John initiated the broadcasting of Newmarket prices through Radio 3DB and was involved in large yardings of sheep and lambs at Newmarket.

On 29 October 1931, John set out with Richard Killeen (also a director at the time) for an auction being held at Oaklands in the Riverina. They travelled together with two auctioneers in a rented car driven by Mr David Tosh. The driver lost control of the vehicle just outside Wangaratta, hitting a tree and overturning the car. Both John McNamara and Richard Killeen were killed.

The Deputy Coroner's Inquest found David Tosh guilty of manslaughter due to "culpable negligence". The case was referred to the Crown Prosecutor who decided not to proceed. The original Coroner's Report has been destroyed and details of the accident obtained from the death certificate.[6]

The fact that both John and Richard were travelling together to the auction strongly suggests that they were looking at the possibility of opening a new branch in Oaklands.

At the time of his death, John was married to Anne Dwyer and left a nine-month old daughter, Marianne. Eventually Anne received only the money John had invested into the company and no other compensation. Richard Killeen's family received nothing.

Patrick McNamara [1845-1922] (Paddy)

Patrick McNamara, known as Paddy, was born in Tent City and was the second son of John and Catherine.

After early life in Tent City and later Springfield, Paddy followed his father and selected 157 acres of heavily timbered land on the

Goulburn River north of Toolamba. He built a log cabin there where he commenced married life with his new bride Catherine Morrissey from Wahring.

Later, Paddy also purchased his brother Michael's selection of land known as "The Lowlands", between McNamara Road and the State Forest.

Paddy began a business in Shepparton as a stock and station agent under the name of J. McNamara and Co. This involved him making a daily ten-mile round trip by buggy on rough roads. Eventually he moved to Shepparton and is said to have owned one of the first cars in that town.

Ten years later, with a family of seven boys (the youngest was Maurice, 16 months old) and two girls, Catherine died suddenly at age 37.

Later Paddy married Ann Smith from Porepunkah and together they had a further five children.

One of Paddy and Ann's sons, Reginald, eventually became the top auctioneer with the firm at Newmarket.

Paddy McNamara had two great friends who often came to the office. Over mugs of whiskey they would discuss the economy, general affairs and old jokes. The three were known as "Father, Son and Holy Ghost".[7]

From his first marriage, Paddy's eldest son William would later become a company director.

Son David followed his father as manager of the Shepparton business and joined with his brother Patrick who controlled Dookie and Tatura and had already established himself when his father died.

The youngest son of Patrick's first marriage was Maurice [1889-1970]. He was 15 years younger than brother David and four years older than Paddy's son Reginald from his second marriage. Mo and Reginald would have been brought up together by Paddy's second wife Ann.

Maurice McNamara [1889-1970]

Maurice McNamara was born in Toolamba and known later in life as Mo or The Boss.

At the age of 14 Mo started work with his father. Later he was to become the managing director of John McNamara & Co. Pty Ltd for 33 years.

Mo was ambitious, a hard manager, and had a reputation for having a great knowledge of stock. He was liked by some, and described as "generous" by at least two people, but others disliked him and feared him for his ruthlessness.

It was said by one stock agent, Arthur Hussey, that if Mo rang, you would only remember half of what he said – but it had better be the right half.

In his retirement interview with a journalist from *The Argus*, Mo claimed that he had purchased John McNamara and Company in Melbourne,[8] however this was not proven by any evidence.

Mo's first marriage, at the age of 19, was to Ada Keiran. Years later, at the age of 75, he was married again, to his secretary Doris Goodge.

There were five children from Mo's marriage to Ada, including Keiran who later followed his father as the Managing Director.

Mo also had a second son Frank who, after some temporary work with his father, joined the RAAF and served as a fighter pilot over Europe. According to Frank (who was well known to me as we were both in the same year during our medical course) he had once crashed his fighter bomber in France but was unhurt.

When he was demobilised, Frank started a medical course at Melbourne University graduating in 1953. He practised as a doctor in Armadale, Victoria. Pat O'Connell recalls that Mo "never spoke of him much".

Keiran McNamara [1913-1981]

Maurice's son, Keiran, was born in Shepparton. He was admired as being a fine man, never seen in work clothes and always well dressed. He was known as the "Beau Brummel of Newmarket", and could apparently be garrulous before a sale.[9]

Keiran married Joy Westlake and had one daughter.

For some time, Keiran was president of the Association of Stock and Station Agents that had been formed following the 1880s wool boom.

When he took over from his father in 1955, Keiran proved to be the last managing director of John McNamara & Co Pty Ltd It is said he did not always get on well with Mo, his father.

John P. McNamara

John P McNamara was the son of Patrick and Catherine, born in Toolamba. He started work with his father in Shepparton and in 1904 moved to Numurkah to work as a partner with his uncle David.

John was highly regarded as a top judge of the value of stock and had a reputation as a "straight goer". He had a close friendship with his and his uncle's business partner Tom Hurley.

In 1897 at the time of Queen Victoria's Jubilee John was chosen as a member of the Victorian Mounted Rifles Troop which took part in the London celebration, and was photographed as a typical Victorian Trooper.[10]

John died aged 50 years.

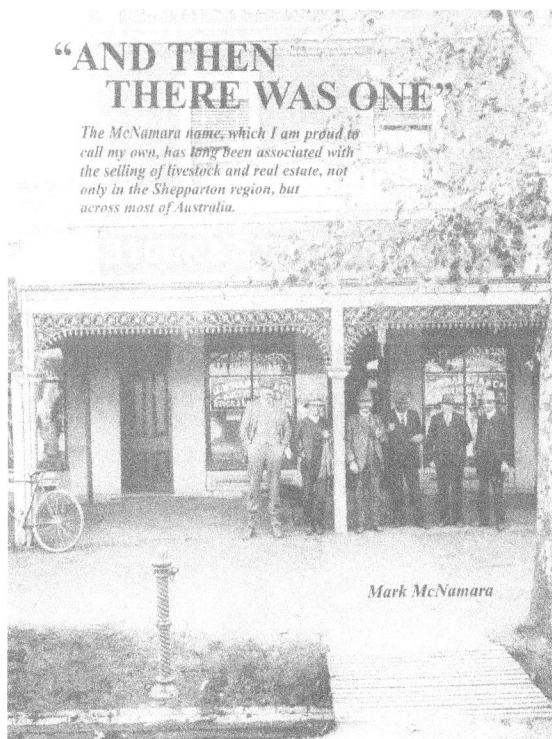

"AND THEN THERE WAS ONE"

The McNamara name, which I am proud to call my own, has long been associated with the selling of livestock and real estate, not only in the Shepparton region, but across most of Australia.

Mark McNamara

The company's Shepparton office
(as seen on the cover of Mark McNamara's book And Then There Was One*)*

Top: Three generations of McNamaras, Maurice, David and Danny, 30 October 1992.
Below: Maurice McNamara & Co, First National Real Estate

David McNamara

David was Patrick's sixth son from his first marriage. He worked

Highlights of John McNamara & Co.

1881   John McNamara founded Stock selling Business in Yarrawonga.

1901   John McNamara conducted Stock Sales at Newmarket.

1908   John McNamara consigned 69 Bales of Wool to England for direct sale, which realised a loss of £413.15s.3d.

1927   13th April. Death of John McNamara, Aged 72 years.

1931   McNamara's submit greatest yarding at Newmarket for all time - 24,000 Sheep and Lambs.

1944   McNamara's Sales at Newmarket exceeded One Million Sheep in that year.

1953   McNamara's sold 7,000 Corriedale Wethers, One Mark A/C Kooba Station in 2 Weekly Sales.

with his father and after Paddy's death became the head auctioneer and manager of the Shepparton branch. He joined his brother Patrick who had offices in Shepparton, Tatura and Dookie.

The McNamara brothers pitching in at Yarrawonga sale, from left, Mick Bill and David

David married Rose Fenaughty and raised a family of eight children, one of whom was another Maurice, known as "Mo Mac". David was also a part owner of the Berry Bros Cordial Factory.

The Shepparton business continued to prosper and was later carried on by David's sons Gerald and Brian. Upon their deaths, as with Yarrawonga, the Shepparton business ceased to exist.

Mark McNamara works in Shepparton in Real Estate with a non-related Company.

Maurice McNamara (Mo Mack) [1928-]

Maurice McNamara, known as "Mo Mack"', was born in 1928. He started with the Shepparton agency at the age of seventeen and later, in 1951, with the agreement of the principals at Melbourne office, was asked to take over management of the Numurkah branch.

After Mo Mack's start in Numurkah he sold a lot of real estate, but commission was refused under orders from Melbourne and the parent company Australian Estates. With this rebuff, he resigned from the company in 1962 and started his own company, operating as Maurice McNamara & Co.

Mo Mack was very successful as an auctioneer, travelling widely and later recalling the story of a clearance sale of 10,000 sheep in pouring rain.[11]

Maurice continued selling real estate and later became associated with First National Real Estate. His son David and grandson Danny have continued the company, which is the last bearing the McNamara name. Maurice retired in 2004.

Formation of John McNamara & Co

With the success of all the brothers' companies, a collective decision was made in 1900 to participate and open a new company at Newmarket Sale Yards and they formed John McNamara & Co.

Individually the family also continued with their own local agencies and in 1916 a total of twenty-one agencies covered Victoria and the Riverina.

References

1. Information from Clare Historical Society.
2. Ian Jones, *A Short Life of Ned Kelly*, Kelly's actual Birth Certificate was not available.
3. Harry Peck, *Memories of a Stockman*.
4. History obtained from son William [Bill].
5. Coroner's Inquest Report in authors possession.
6. Original Coroner's Report has been destroyed-details of accident obtained from Death Certificate.
7. Mark McNamara, *And Then There Was One*.
8. *Argus* journalist report on his retirement as Company Manager.
9. Recalled by Jim Drake.
10. Obituary notice in Numurkah paper.
11. Discussion with Maurice McNamara (Mo Mack) in Numurkah.

PART THREE

The Newmarket Years

3

Livestock sales in Melbourne

As Melbourne grew in the first 26 years, livestock was sold at various sites around the city, including Kirk's Bazaar in Bourke Street, which specialised in horses. Some sales also took place in hotel yards, and Flagstaff Hill was a site for cattle sales.

The main site for the sale of cattle and sheep at the time was at the corner of Bourke and Queen Streets. Drovers would drive livestock through the city streets – a practice that was seen in living memory when Jim Drake witnessed a mob of cattle driven down Spencer Street to the South Melbourne Market.

The first really organised livestock market was at the corner of Elizabeth and Victoria Streets, close to butchers in the Victoria Market.

Establishment of Newmarket Saleyards

In April 1849, the Victoria Racing Club offered to give land adjacent to the saltwater river (Maribyrnong River) for an abattoir and cattle pens. This became the site of the Newmarket Saleyards and the city abattoir.

The complex was opened in 1861, with nineteen stock agents, and operated on this site for 126 years before officially closing in July 1985.

In 1894 it was decided by all agents that sales would be by public auction and in 1897 the Melbourne City Council established rates. At that time Harry Peck was granted an auctioneer's licence which he went on to hold for 54 years.

After the first public auction was held in 1894 the popularity and significance of the market rapidly increased.

Transportation of stock on Victoria's railways

The first railways in Victoria were opened in 1854. It was 1861 when the railways first brought sheep from Geelong to Newmarket. Subsequently drovers and their dogs had only to drive livestock from the Flemington railyards to the Newmarket pens.

On one occasion, years later in 1952, of the one thousand railway trucks in the yard, 385 of those contained livestock destined for sale by McNamara's.

Establishment of John McNamara & Co

It was the rapid growth of Newmarket that attracted the three brothers John, David and Patrick who until that time had been operating their businesses independently. They now joined together to move to Melbourne as John McNamara and Co opening at Newmarket in 1901, and with an office also in the city in Melbourne.

Newmarket became a place of great excitement and challenge for all who worked there, and there are many happy memories shared by those who spent their time in that location.

In 1908 the company ventured from livestock sales and sent a large consignment of wool to England on their own account. £413 was lost on the deal, and so dealing in wool was not to be repeated.

The sale by John McNamara & Co. Pty. Ltd. at Newmarket, 19 December 1944, which exceeded one million sheep for the year

A typical day at the Newmarket Saleyards

The day at Newmarket started before dawn with drovers moving livestock from the railyards to the big pens. All were very hard workers and many were also hard drinkers.

With the help of the dogs, agents of McNamara's then arrived to move their clients' stock to the sale yards. George Lang was the supervisor of this drafting and made it look easy. All was colour and movement – appearing as organised chaos. Mo, The Boss, although a very particular man, apparently never took exception to George Lang's work.

The next activity of the day was the draw for order of sale, a draw that was taken from a wooden barrel.

The auctioneer, together with his clerk and drover, would go to the sales pen where sales would begin at 8.10 am after the ringing of the large bell.

The Newmarket staff, 1944: Pat McNamara (2nd left back row), Reg McNamara (2nd left front row), next to him, Kieran and Maurice, and Brian, far right

The auction would start with the auctioneer standing on the pen calling for bids from the buyers standing around. The auctioneer was an excellent judge of the value of stock and what price it should bring.

On one occasion, at the sale of stud rams from a valued client of McNamara's, the heavens opened and Mo is said to have ordered the rams under cover for the night, so they were moved into the Company office.

It was a great innovation that took place in 1930 when John McNamara (1892-1931) arranged with Radio 3DB to broadcast daily stock sale prices at Newmarket. This service informed farmers, allowing them to know the current value of their stock.

Following McNamara's initiation of this broadcast, the same service was later taken up by other country stations and the Australian Broadcasting Corporation (ABC).

Auctioneers and other staff at John McNamara and Co

During the 1930s and 1940s McNamara's had a large staff with top auctioneers including Paddy's son Reginald McNamara, who was an expert with store sheep. Reg would travel widely to pick top stock for sale, but although he was Mo's stepbrother he was never made either a partner or director of the firm.

Another of the early auctioneers at Newmarket was Mo's uncle, James McNamara (son of John and Catherine), who worked as an auctioneer for another firm, Watt's Turnbull & Company.

Keith Vincent, author of the classic book *The Fall of The*

Marianne Brenan (nee McNamara), John's daughter, at Newmarket

Images taken at Newmarket in recent years
Top: The front of the Newmarket sales office. Bottom: Newmarket Clock Tower

Hammer, is another who worked for the McNamaras for many years as a livestock man.

Joe and Mick Brannigan also provided strong yard support for the McNamaras during those years.

General references used throughout this chapter

Keith Vincent, *On The Fall of The Hammer*, State Library of Victoria, 1993. (The classic book written on the history of Newmarket).

Discussions with Jim Drake.

Discussion with the late Peter McNamara.

PART FOUR

Company Governance

PART II

4

Company Governance

Although John McNamara and Co was a great success, by 1922 all three brothers, Patrick, David and John, were diagnosed with cancer[1] and it became obvious that changes would have to be made to the structure of the company.

The following advertisement appeared in *The Argus* (7 September 1921):

> John McNamara Stock and Station Agent of 443 Bourke Street, Melbourne, wish to make known that their old established auctioneering business has been formed into a Proprietary Company under the name of John McNamara and Company Pty Ltd, and the following additional members have now joined the company and will take an active part in conducting future business:
>
> > William McNamara of Tatura,
> > Maurice McNamara of Shepparton and Dookie,
> > John McNamara Jnr of Melbourne,
> > Richard Killeen of Melbourne.[2]

William and Maurice were both sons of Patrick (Paddy), with William being seven years older than Maurice. Their brother David was the manager of the Shepparton Branch.

NEWMARKET FAT STOCK SALES.
NOTICE
To Our Clients and the Public Generally.
JOHN McNAMARA and Co.,
Stock and Station Agents,
443 BOURKE STREET, MELBOURNE.
Wish to make known that their Old-established
Auctioneering Business has been formed into a pro-
prietary company, as from the first day of Septem-
ber, under the name of
JOHN McNAMARA and Co. PROPRIETARY
LIMITED,
and the following Additional Members have now
joined the company, and will take an active part
in the conducting of future business:—
WILLIAM McNAMARA, Tatura
MAURICE McNAMARA, Shepparton and
Dookie.
JOHN McNAMARA, Jun., Melbourne
RICHARD KILLEEN, Melbourne.
By strict and personal attention to all details
essential to the successful marketing of stock, the
newly-formed company hope to further increase
their list of already numerous and esteemed
clients.
FOR TRUCK ORDERS WIRE
McNAMARA Co., MELBOURNE.
'Phone 4245.   —(Advt.)

John McNamara and Co Pty Ltd, under new management

At the company's formation there was no indication as to who was to be Managing Director. It would seem that John was the logical choice due to his successful running of the Melbourne office and Newmarket for six years from 1916. However, there was no mention of who should be first among equals and Maurice (Mo) became Managing Director – possibly through election, but more likely by the force of his personality. It does appear that he imposed himself and became the SAM (self-appointed manager).

From this time on, Mo was always referred to as "The Boss". He was now the Managing Director of a well-established business at Newmarket with twenty-one country agencies.

Upon his retirement Mo claimed that he had purchased the business, however there is no evidence at all to support this statement.[3]

It was not long after the formation of the new company that all three brothers passed away. Patrick died in 1922, David in 1923 and John – who had virtually retired in 1912 – died in 1927.

Financial problems for the company

An early problem arose during the depression when banks would not lend money to farmers. The farmers therefore relied on Stock and Station Agents to waive their sales commissions and to allow them to borrow agency money.

To save the agricultural industry after the depression, an Act of Parliament was passed cancelling farmers' debts. This created financial problems for John McNamara & Co Pty Ltd and led to groundless rumours of company foreclosure by the banks due to debts.

It is suggested that Mo pointed out to the banks that many farms would be destroyed and they would get nothing, at which the banks relented.

On one occasion Mo was accused of underselling livestock by a few sheep. It was almost certainly a mistake but Mo, defensive of his reputation, threatened to sue for defamation. The matter was settled out of court with an apology.

The early days of John McNamara & Co Pty Ltd

Under Mo's management, the company flourished and by 1930 had the biggest turnover at the Newmarket sheep section. In 1931, before John McNamara's death, the greatest yarding for that time was over 24,000 sheep and lambs.

Although Mo could be ruthless, he was a shrewd judge of both stock and men. He established a large company at Newmarket with top auctioneers Reg McNamara, Tom Guinan and others. One of these others was George Clarke, a fat lambs auctioneer, who after a fallout with Mo then went to Goldsborough Mort.

Fred Nathan was responsible for handling all the sheep and cattle from dealers in New South Wales, Queensland and South Australia, while Jack Coglan was head of all rail loading which, before road transport, was an enormous job. Jack would be on the telephone every day of the week to branch managers all over Australia.

Tom Guinan was a top auctioneer of fat sheep and cattle. He was auctioneering one day, standing on the pen, when he suddenly collapsed and died. Like show business where "the show must go on", auctioneering went on, hardly missing a beat.

There were only two occasions in recent times when auctioneering did stop at Newmarket: the first, when prisoners of war returned to Newmarket, and the second, when Australia won the America's Cup.

A change in company management

The tragic deaths of John McNamara and Richard Killeen, when they were travelling to investigate the possible opening of a branch in Oaklands, left only the two brothers, Mo and his elder brother William, as sole company directors.

This should have been a golden opportunity for Mo to make his brother Reginald a director. Reg was universally liked and regarded as the top auctioneer at Newmarket and had earlier come to Melbourne from Shepparton with an expectation of being a partner or director.[4] However this did not happen.

In 1933 William had a disagreement with Mo, and left the company to work in Queensland, where he became a financial success as a stock agent. He later returned to Finley in New South Wales where he purchased a property.[5]

Mo closed ranks at that time, appointing his son Keiran a director and his brother-in-law James Francis Keiran (known as JFK) as the company accountant. Mo became the face of the company, retaining complete control.[6]

Later there were also suggestions of friction between Mo and his son Keiran.

Growth of the company

Under Mo's leadership the company grew to become the largest stock agency in Victoria and the Riverina. Together with twenty-one country agencies, there were a large number of pocket-book agencies consigning stock from country railway stations to Newmarket.

A large interstate store stock connection was established throughout New South Wales, Queensland and South Australia.

Although the attempt to open an agency in Oaklands failed when John and Richard died in the car tragedy, later in the

1940s five franchise companies were established in Swan Hill, Echuca, Kyneton, Wodonga, Moulamein and Deniliquin with the outstanding agent Arthur Hussey.

In 1944 the company created a world record auctioning 1,019,018 sheep in a single year, a Commonwealth record auctioning 33,954 cattle in the same year, and set another Commonwealth record for auctioning 6,000 sheep in one day.

Australia was certainly riding on the sheep's back. The commission for these sales has been estimated at £1,000,000.

With Mo as Managing Director, son Keiran as a director, and Frank Keiran as accountant, the company appeared to be thriving.

John Francis Keiran (JFK)

The company accountant, John Francis Keiran (also known as Frank or JFK), lived in a luxurious second floor apartment at the Alexander Hotel (Savoy Plaza) in Spencer Street. Frank had separated from his wife who was living at the expensive Cliveden Mansions in East Melbourne.

Pat O'Connell recalled that while he was a junior with McNamara's, one of his duties was to deliver a carton of Craven A cigarettes every one or two weeks to JFK. He would sometimes bring cheques for signature, and was able to personally observe the luxury of the company accountant's apartment.

Pat also delivered cigarettes to Frank's wife who he felt "liked a drink". He recalls too that JFK apparently drove a maroon Buick or Pontiac.[7]

There had been no serious problems with the management of

the company until in 1948 it was discovered that John McNamara and Company Pty Ltd was in grave financial trouble. It had not been realised earlier that JFK was not auditing the accounts and that in fact he was a gambler.

JFK's losses on the stock exchange had virtually bankrupted John McNamara & Company Pty Ltd. When this came to light, it appears that Mo had a nervous breakdown.[8]

Sale of the company to Australian Estates

To recover, Mo travelled to America with his daughter Patricia. In America, Patricia met and married Jack Little who later came to Australia as a radio and television presenter of wrestling and boxing.

More importantly, Mo then travelled on to England and organised the sale of the company to Australian Estates, a 19th century English company incorporated in 1894, which had commenced business in Australia in 1936.

Australian Estates was a huge, diversified company with extensive interests in the sugar industry, livestock, properties and stock and station agencies all over the country and an interest in acquiring more in order to operate Australia-wide.

Mo negotiated the sale of John McNamara and Company Pty Ltd himself, holding no discussions with any other interested parties. Subsequently Australian Estates conducted negotiations individually to purchase all the company agencies.

When Mo returned to Melbourne no one had any knowledge of the company sale.[9] The main stay of Newmarket auctioneers, including Reg McNamara, Jack Coglan and Tom Guinan were furious to learn all they received for their lifetime of service was £4000.[10] Shortly after the takeover, although Mo was still in control, Reg retired from auctioneering and worked part time in administration with Fred Nathan.[11]

Sydney Morning Herald

300-FT FALL TO DEATH
Melbourne Man Had Been Ill

A Melbourne man yesterday handed 35/ to a taxi-driver who had driven him from Hurstville and then fell 300 feet to his death at Jacob's Ladder, Watson's Bay.

He was James Francis Keiran, 62, retired, of Toorak, Melbourne.

Police were told that he had been ill and was discharged from a Sydney private hospital a few days ago.

The taxi driver, David Edward Smith, of Greenacre Road, Hurstville, said Keiran engaged him at Hurstville and said he wanted to go to Watson's Bay for a drive.

He asked him to stop at Jacob's Ladder, because he wanted to enjoy the sea breeze.

"As we were walking towards the safety fence he asked me to take his spectacles back to the cab," Smith said.

"I looked back and saw him climb through the fence. I shouted, but he disappeared."

Police used cliff rescue gear to recover the body.

More tragedy

Seemingly the due diligence carried out by Australian Estates revealed Frank Keiran's malfeasance. He became very depressed and travelled to Sydney where he spent three weeks in Lewisham Hospital.

On Friday, 28 April 1950, Frank discharged himself from hospital, engaged a taxi in Hurstville, and asked to be driven to Watson's Bay. At Jacobs Ladder, at the Gap, he asked the taxi driver, David Smith, to stop so that he could enjoy the sea breeze. As they walked towards the safety fence, JFK asked Smith to take his spectacles back to the taxi. "I looked back and saw him climb the

INQUIRY TOUCHING THE DEATH OF JAMES FRANCIS KEIRAN '

I FIND that the said JAMES FRANCIS KEIRAN, on the
twenty-eighth day of April, 1950, at the foot of
cliffs at Watson's Bay, in the Metropolitan Police
District, in the said State, died from injuries to
the head and body, wilfully self inflicted by jumping
from the top of such cliffs, at the same place on the
same date.

SLM.
CITY CORONER.

AA
CITY CORONER'S COURT, SYDNEY,
26TH MAY, 1950.

fence, I shouted, but he jumped and disappeared" recalled Smith. Police later retrieved the body from the bottom of the Gap.[12]

It would appear that Frank became depressed upon realising the part he played in the company's destruction, but we will never know if he also had an underlying psychiatric problem such as bipolar disease, as all records of the now closed Lewisham Hospital have been destroyed.

He left a Will, dated January 1950, leaving assets of £6,734. One listed asset was:

600 10/- shares in Nell Gwynne Reef, valued at sixpence each and realised £15.[13]

Management of the company under Australian Estates

When Australian Estates purchased the company and all its agencies, they allowed John McNamara and Company Pty Ltd to continue its management without direct control, however they did fully control the finances. This is demonstrated by the annual financial report of 1951 and 1952 in my possession (see Appendix 2).[14]

In 1955, Mo retired and company management passed to his son Keiran.

Then in 1968 John McNamara and Company Pty Ltd was completely absorbed into Australian Estates. The residual management of the assets was controlled for Australian Estates by Brian Rodwell, who does not recall seeing any Minutes of Directors Meetings.

Shortly after, through mergers and takeovers, Australian Estates disappeared and all records passed to Elders and have now been lost or destroyed.[15]

Finale

Although Mo supervised the significant growth of the company during his 33 years as "The Boss", it is clear that he failed in some aspects of the governance of the company.[16]

The fact that the company went from riches to rags between 1944-1948 suggests that it was under Mo's management and

The Argus (Melbourne, Vic

52 YEARS ON JOB

Mr. Maurice McNamara, one of Victoria's best-known stock and station agents, retired yesterday as managing director of John McNamara and Co. Pty. Ltd.

Mr. McNamara held the position for 33 years.

He has been connected with the stock and station agent business for 52 years.

In 1921 he sold out his interest in J. McNamara and Co., Shepparton, and purchased the Melbourne firm of John McNamara.

His company sold more 1 million sheep through Newmarket in 1944.

This still stands as a record for any central market in the Commonwealth.

His eldest son, Mr. Keiran McNamara, fills the vacancy.

supervision that this iconic company was destroyed. Perhaps he was 'asleep at the wheel' or for whatever reason failed to confront his wife's brother. In any case, he left the finances in the hands of JFK, who never arranged the yearly audits that would have saved the company. Mo consulted nobody when he finally realised what was happening, and went on to sell the company to Australian Estates, all in secrecy.

It was an era, however, where many large Australia-wide companies were being merged or taken over, and Australian Estates and Elders were the assumed future in the industry.

After Mo's retirement in 1955 he continued independent trading in cattle for a further five years, continuing to cause fear at Newmarket and always demanding the best sales pens.

Following the closure of the original company in 1968, Michael, William [Bill] and David, all sons of John Joseph McNamara [1888-1967], along with other McNamara descendants, established agencies in Finley New South Wales, Numurkah (Maurice), Deniliquin and in Melbourne and Shepparton (Keiran McNamara).

Michael McNamara continued in Yarrawonga until his death in August 2007.

The glory days of this iconic stock and station firm were over however, leaving only Maurice McNamara, his son David and grandson Danny to continue the family tradition in Numurkah.

References

1. Death Certificate Information.

2. *The Argus*, 7 September 1921.

3. "Fifty Two Years on the Job", *The Argus*, 21 December 1955, p. 9.

4. Discussion with the late Peter McNamara.

5. Discussion with Pat O'Connell.

6. Ibid.

7. Ibid.

8. Ibid.

9. Discussion with the late Peter McNamara.

10. Discussion with Pat O'Connell.

11. Discussion with the late Peter McNamara.

12. Coroner's Report held by author John A.Brenan. Police Report. Report, *Sydney Morning Herald*, 29 April 1950, p. 5.

13. Will held by author John A.Brenan.

14. Financial statement of Australian Estates in records held by A.N.U.

15. Discussion with Brian Rodwell.

16. Discussion with Richard Killeen's nephew.

Sources

Garden, Don, *Victoria – A History*, Thomas Nelson, 1984.

Harris, Helen Doxford O.A.M., Research Historian.

Information from discussions with the following:-

1. The late Peter McNamara.

2. Maurice McNamara (Numurkah).

3. William (Bill) McNamara.

4. Pat O'Connell.

5. Jim Drake.

6. Brian Rodwell.

7. Margaret Burke.

8. Mona Kelly.

9. John McNamara.

10. Mark McNamara.

Irish history publications listed as "References" to Part One.

Marriage and Death Certificates from Victoria and Irish Registries.

McNamara, Mark, *And Then There Was One*.

Vincent, Keith, *On The Fall Of The Hammer*.

Wills and Coroners' Reports.

MEMOIR OF HIS EARLY LIFE

Written by Maurice (Mo) McNamara

Maurice (Mo) McNamara's Memoir

"The Lowlands", Toolamba, was the place of my birth on 19 October 1889. My parents, Patrick and Catherine McNamara, had me Christened at the Catholic Church, Mooroopna (eight miles away). They called me Maurice.

My father was the son of Irish immigrants from the counties of Cork and Clare, but he was born in 1845 in Canvas Town (Melbourne), somewhere around the area of old Kirk's Horse Bazaar in Bourke Street.

When my father was just a lad, the family moved to the Springfield district in the Romsey-Kilmore area and later settled in the Goulburn Valley handy to Mooroopna.

As a young man my father (one of a very large family) along with his brothers, Michael and John, sought distant fields to make way for the younger members of the family on their limited holding. They eventually found work in the outback country some two hundred miles west of Hay when the well known early pioneer, the late Mr Peter Tyson, set them to the task of building a "Turn-back Jimmy", which story greatly enthralled me as a youngster. I later learnt this was the banking up of a depression – by means then of the old bullock teams – to conserve large holdings of water for stock to carry over the dry periods and droughts which, from hearsay, were so frequent in those times of early settlement.

While these young men were making their way their parents' Mooroopna land was sold and other members of the McNamara family spread throughout the Goulburn Valley.

My father eventually took up a river selection on the Goulburn River some three miles from Toolamba in rather low lying and heavily timbered country. Michael went to Cobram and John to Yarrawonga where he founded the well known stock auctioneering firm of John McNamara & Company around 1880.

The younger brothers, James, Maurice and David, all later followed the auctioneering profession in Queensland, along the Murray Valley at Cobram and Numurkah as did my father, Patrick, about the same time at Shepparton.

My father built a log dwelling on his river selection at Toolamba. This was constructed of hand-hewn timber from the property and comprised two or three rooms. A few years later he took to his home his bride, Catherine Morrissey, of Wahring.

"The Lowlands", adjoining the railway station at Toolamba, was later purchased and most of the family of seven boys and two girls (Jack, Thomas, Michael, Patrick, William, David, Catherine, Margaret and Maurice) was born there. My mother died during the former part of 1891 when I was sixteen months old.

In the early eighties, in addition to conducting his mixed farm at "The Lowlands", my father commenced his auctioneering business at Shepparton and invariably drove in his hooded buggy and pair or the jinker ten miles over very bad and flooded country to his business and home each day.

I understand that it was not unusual for my father to be accompanied on his return at 7 or 8 o'clock in the evening with two or three of his clients all of whom would enjoy a hearty meal – prepared it seems at a minute's notice. They were sometimes also bedded down for the night – the members of the family "doubling up" on such occasions. Such was hospitality in those days. What heroines the women were – looking after large families and visitors, most times without help except from the older children of the family.

Amongst my very early memories are the journeys in our double-seated open buggy which left five mornings of the week well loaded with members of the family (after a big dairy herd was milked) to travel five or six miles to school on the main Tooloomba-Mooroopna road.

My father had a great love and knowledge of horses and was a game and daring driver as many a friendly traveller knew to his regret when he would try to pass Dad on the limited road space available. How proud was I to sit beside him when our buggy and pair had raced first all the way and went full-tilt into the church-yard to the amusement of many of the parishioners. The talking over the events of the week by the adults after Mass seemed to me to last an eternity.

When later we lived in Corio Street, Shepparton, it was ritual to call on Grandmother Catherine Gorman McNamara. She was a tall lady who seemed always to be dressed in black and, although I found her ghost stories very enthralling, I almost feel the chill down my spine now as I did then as a lad when her description of her various encounters with ghosts was so graphic. I hear her now say, "My name be Catherine Gorman, I'd say", in answer to the ghost's query.

My father remarried in 1893 to Miss Anne Smith of "The Uplands", Porepunkah, where her people had a selection just at the foot of Mt Buffalo. Dad had always worn a beard but, apparently for his nuptials, had it shaved and how well I remember my first encounter with this "clean shaven stranger". It had always been a great joy to me to meet Dad whenever he returned home by train from Melbourne and to trot by his side across our paddock to the homestead. On this evening the Toolamba Station was, to me, as usual the most magnificent edifice in the world with its gleaming kerosene lamps enhanced as they were with a backing of aluminium reflectors, but the beardless stranger who smilingly approached me was not my father I thought and I scampered home without him as fast as my four-year-old legs would carry me.

With the passing of years some of my older brothers left home and farm duties – John to enter the auctioneering business with our Uncle David at Numurkah, Patrick to manage the Tatura Office and William joined the Shepparton staff as auctioneer.

By the time I was of age to be educated a school had been built at Toolamba West and to this I trudged to and fro 1½ miles each day, but in the mornings not before I had gathered the "morning's wood", fed the poultry, calves and pigs and helped with the milking of some seventy cows. Similar duties were expected of me on return from school in the afternoon.

One large room was the style for country schools at my time of learning as it has been for many years subsequently, and the teacher coped with the instruction of pupils at all class levels. As I talk of school I particularly recall one teacher, Mrs Lane, standing at the door on hot days flicking the flies from the backs of pupils, with a towel, as they filed into the classroom. No pressure pack sprays in those days! I proudly recall, too, that the McNamara family was renowned for regular attendance at school.

School over for the day and my share of farm tasks done I looked forward to the return of my father from his day's work when he would recount to us the happenings in the auctioneer world. Dad was a great reconteur and in later years when I, too, joined the business and travelled with him and other to sales by train – as was the pattern in those days – I greatly enjoyed his story-telling and his enthusiasm when playing cards. I can see him now thumping the table with glee when he had scored a point.

My recollection is of full and plenty at "The Lowlands". The butcher called once a week and we killed a lot of our own stock and, throughout the year, many pigs for bacon. I can see now the half sides of bacon hanging from the rafters. It was a rare day on the farm when a pig was to be slaughtered – all and sundry taking part in the activities from the time the pig was eventually lassoed in the open yard after much chasing until it was hauled up to a post to be "stuck", attempts being made to catch the poor thing's blood in a dish for the making of black puddings. Then into a 50 gallon boiler of water with the dead animal, the water having been brought to boiling point with the boiler placed on rails over a log fire. After a short immersion the pig would be hauled out – water splashing – kids scampering – and Dad shouting orders for all hands to commence de-hairing of the animal. What a circus! Sometimes the pig had to be returned to the boiling water if its hair did not come off easily and, of course, the hullabaloo would be repeated. For many days following this event we enjoyed succulent dishes of pig cheeks and trotters and watched the making of many pots of ointment from the lard. This ointment had a reputation second to none as a cure for all sorts of sores and cuts.

After some years my father, step-mother and the second family of one boy and four girls (Reginald, Anne, Irene, Gladys and Mollie) took up residence in Shepparton but I stayed on with my brothers at the farm under the care of a housekeeper. A little later, however, I joined the family in Shepparton and attended the Convent of Mercy Day School.

When I was about twelve years old I was sent to a school with the important sounding name of Hassett's University School. The classes, incidentally, in the early stages were conducted at the rear of the teacher's home in Corio Street in an old stable into which a wooden floor had been built some feet above ground level. Sometimes for divertissement and sometimes accidentally my pencil would fall through the opening in the floor boards and after school I would scrounge around to locate it in the darkness of the manure covered ground of the stable below.

Mr P.A. Hassett, the Principal, was a man of great integrity and a wonderful teacher who also conducted night school at that time in Shepparton. I remember an incident on April Fool's Day when my pal (Jack Thompson, son of the local chemist) and I decided to give the night school boys a scare by creeping stealthily up to the school – a short distance from our home – and igniting a giant size cracker on the doorstep. We scampered back post-haste to crouch behind our high front hedge to observe the consequence of our deed but, after the explosion, quailed inwardly when we espied Mr Hassett standing in the doorway brandishing a menacing stick in his hand. From comments at school next day someone else had stood the blame, Mr Hassett being reported to have said some boys he found near Kitto's Dye Works wouldn't be throwing crackers in the school grounds again in a hurry.

Hassett's college was later conducted from more pretentious premises in a little street at the back of the old Union Hotel in Fryer Street. Some years after this Mr Hassett – who had prior to teaching been on the clerical staff in the business of John McNamara, Yarrawonga – commenced teaching at Prahran where he built up a large Commercial College. At this time his brother was a well known solicitor in Melbourne.

I finished my formal schooling at the age of fourteen having told my father I did not wish to go to Xavier College in Melbourne as my older brothers had done.

I have to confess playing the wag from school now and again to go to the saleyards where I found the bustle and excitement of the stock sales very fascinating. After school on many occasions my pals and I would dash down to the saleyards and I would stand on the fence emulating the auctioneers I had so often watched while my young friends would bid for the imaginary stock. No wonder then that I entered my father's business on leaving school.

For three years I was "Jack of all trades" under tuition from Dad, his very well know Manager of many years – John McInerney – and my brother David, who was then the Chief Auctioneer for the firm, brother William having gone to Queensland where he sold for the large firm of McPhee & Co, Toowoomba. He later opened his own business at Oakey on the Darling Downs. I will relate more of my association in business with William at a later time.

I might mention here that Mr Jack McInerney had, before joining my father, been in partnership with my Uncle James at Cobram in an auctioneering business under the name of McNamara, McInerney & Co I do not know for how long nor with what success the business was conducted.

Earlier my father's eldest brother, Michael, had conducted a Stock Auctioneering Business at Cobram and continued in it for many years with his large family of sons including James (later of Berrigan) who for many years auctioned for Watts, Turnbull & Co at Newmarket. This Cobram family also could boast a famous son in David who played outstanding football for many years with St Kilda and Essendon. Until the time of his death in 1968 he still held the World Record for the Longest Place Kick – some ninety odd yards. Dave was a wonderfully well-built fellow of 6'4" in height. He told me his greatest asset as a footballer was the fact that he could stretch almost the same width as his height; consequently he could mark over any opponent. Dave held a record of kicking eighteen goals in a match played in Melbourne and just prior to his death showed me this very football which he used on that occasion. It was mounted on an oak stand and stood in pride of place on the mantelpiece in the sittingroom of his home in Glen Eira Road, Caulfield. The ball was still in good shape but rattled when shaken as the rubber bladder had perished as well it might forty-five years after the great event. I understand Dave's daughter intended presenting this to the St Kilda Football Club.

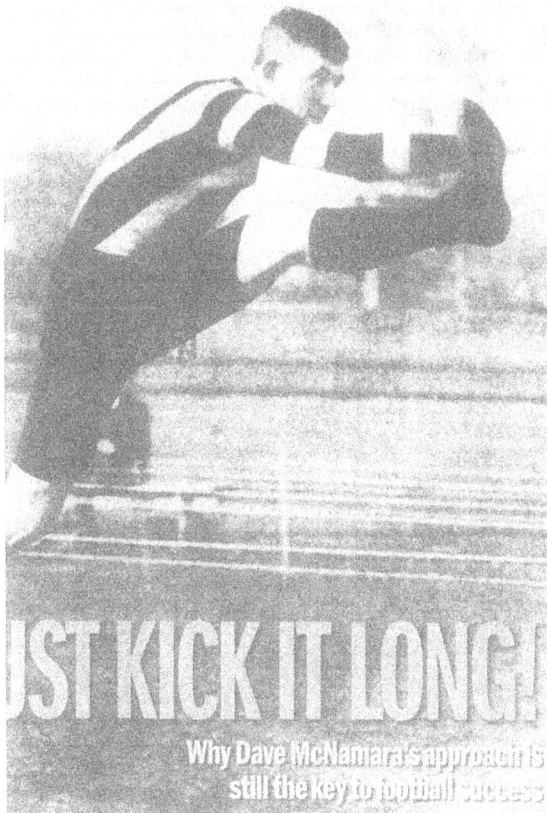

JST KICK IT LONG!
Why Dave McNamara's approach is
still the key to football success

When I was seventeen years old my father decided to build saleyards at Dookie – the only country with some undulation in the Goulburn Valley – approximately 18 miles from Shepparton. With initiative or the desire to launch out on my own I had the audacity to ask Dad who was going to manage the new business. He replied with what I wanted to hear but still to my astonishment, "You can have a go if you think you are able". Two years later he showed the same big-mindedness and faith in my judgement when I rather timidly told him, as we descended the hill into Dookie in our gig, that, young as I was, I wished to marry Ada Keiran.

I was installed at Dookie as Manager Supreme of all I surveyed which was the brand new sheep and cattle yards, a horse and gig and an office for which I paid the magnificent sum of 7/6d (75c) rental per week. My first solo effort was the finding of board and lodging and this I did at the Dookie Hotel under the care of Mr and Mrs Thomas Bollard.

Appendices

Appendix 1

VX. 30994 Lieut. McNAMARA, John Francis.
- 2/24 Aust. Inf. Bn.
AWARDED M.C.
CITATION OF AWARD.

For outstanding gallantry and devotion to duty. On the morning of 10th July, 1942, during the advance on TEL EL EISA RIDGE when the battalion went into the attack, this Officer had command of the Platoon on the right flank whose duty it was to clear the sand dunes of enemy machine gun nests.

During this action he personally led his men with outstanding gallantry and although the enemy were well established he overcame all resistance and the right flank of his Battalion as it advanced to the attack was secure.

When the enemy attacked in force on 12th July, 1942, this officer was again in command of the right flank on the sand dunes on the sea coast. Again with conspicuous gallantry and complete disregard for his own personal safety he inspired his men and when no other was available, he personally controlled the fire of 3" mortar on his flank. Again by his inspiring example, gallantry and devotion to duty and thought the enemy made a determined thrust along the sand dunes, the position was held. On yet a third occasion on 13th July this officer behaved in exactly the same manner. This time he was wounded but his position was held.

Description of the Battle where
John F McNamara won M.C.

At Zero Hour 'B' Company moved forward under the able leadership of Lieutenant J F McNamara. Herb Griffiths is just about the proudest man in the Eighth Army, as he is the right hand marker for the whole push. The Arty are pounding the enemy wire and his foremost defended localities for twenty minutes and then are lifting the barrage in bounds, so we have high hopes that when we get there, Jerry will be so bomb happy that there won't be much fight left in them. Soon after crossing the little white tape, Jerry's mortars opened up on us and Lieutenant Pop Hetherington, commanding 12 Platoon, was hit and killed instantly. Sergeant Henry French assumed command and 12 Platoon seemed to be having a bad time. Big Mac was hit just before we reached the wire, but, after slapping a bandage on, took over the command again.

Meantime, at 10.23pm, 'A' Company, under command of Major Bill Mollard, and 'C' Company, under command of Captain Ted Harty, had moved from the start line with 'A' Company on the right, and Bill Mollard recalls that as his company passed through 'B' Company he noticed "Big Mac", that mighty soldier, with several bandages on his neck, arms and leg, hopping around, using a rifle as a crutch as he directed the 'B' Company action...

...The following day passed quietly with intermittent shelling. Lieutenant Jack McNamara, who had been wounded in four places in the initial attack, was ordered out by Colonel Weir and Lieutenant Fred Geale took over command of 'B' Company.

At 7pm Colonel Weir called a conference of the 'O' Group and issued his orders for a further attack to commence at twenty minutes after midnight on 26 October.

The unit had to move west some 200-300 yards and form up on a start line in the 2/48th Battalion area, from which it would attack the enemy defences from the flank, attacking north-east for a distance of 3000 yards to the Fig Garden area.

With 'A' Company on the right and 'B' Company on the left, the battalion would attack on a two-company front, moving off forty minutes after the 2/48th Battalion, which would attack due north towards Point 29, the highest feature in the area. On reaching a point half way to the Fig Garden, the leading companies would allow 'C' and 'D' Companies to pass through to capture the final objective.

Forty minutes after midnight on 26 October, the leading companies crossed the start line. The enemy had, of course, had ample warning of our approach, as the 2/48th Battalion was already heavily engaged on our left.

Description of the New Guinea Battle where John F McNamara was killed

Meanwhile 'A' Company was ordered to pass through the 2/23rd exploit to the Wareo-Bonga track junction. 'C' Company was to follow and 'D' Company to cut the Wareo-Bazuluo track. Soon after 11am the first enemy delaying force was contacted, "Snowy" Yarwood, a forward scout from 7 Platoon, being mortally wounded.

He was a fine solder and a great loss to his platoon. A flanking move was made by 9 Platoon commanded by Lieutenant Gerry Stretch; the position was cleared and the company continued to advance, reaching the track junction at 3pm, only to find it covered by concentrated enemy fire from the "2200" feature, the crucial ground in this area. 'A' and 'C' Companies formed a defensive area and tried unsuccessfully to find a route through the precipitous country on the northern flank.

Meanwhile, west of Wareo, the battalion had met with serious opposition. After cutting the Wareo-Bazuluo track, Captain McNamara, with 'D' Company, was ordered to occupy Bazuluo. At 1.50pm the rapidly advancing company was ambushed and the leading platoon was cut to pieces, but those remaining extricated themselves, five of them being wounded.

Here Sergeant Peter Atkinson showed magnificent courage in going forward to the ambush area and, although under continuous fire, carried three wounded to safety and returned a fourth time to find the remaining dead. Initially wounded, they had been killed by the heavy fire the enemy kept pouring into the area. There were still eight men missing, including Captain McNamara.

Lieutenant Halliday took command of 'D' Company. A company from the 2/23rd came to assist and sent a platoon to encircle the enemy. A section patrol from 'D' Company found the bodies of the missing. The total loss through the ambus was eight men killed, including "Big Mac", and eight wounded, a very severe blow to the battalion at this stage of the campaign.

The death of Jack McNamara, who was a legendary figure in our Division, cast a cloud over the closing stages of the campaign. He was typical of the superb leaders who commanded in the field at this time, and whose unfailing courage, leadership and enthusiasm tempered with experience and skill played a great part in making Australia's jungle army such a formidable fighting machine. Captain McNamara was a born leader, a man of kindness, courage and riotous good humour, who inspired his men to encompass seemingly impossible tasks by his own total disregard of danger.

"Big Mac" had always ordered that if he were killed in action he must be buried with his boots on. When his body was brought in his boots had been removed by the Japanese; this worried his company to such a degree that an unofficial patrol set out to recover them, returning at dawn to quietly carry out their company commander's orders.

Appendix 2

JOHN McNAMARA
CONSOLIDATED BALANCE SHEET

£. s. d.

LIABILITIES (CONTD.)

Brought Forward –   443,556 19 10

We have examined the above Balance Sheet and
annexed Profit and Loss Account of the Company which are
in agreement with the Books of Account and returns. We
have obtained all the information and explanations which
to the best of our knowledge and belief were necessary for
the purposes of our audit. In our opinion proper books of
account have been kept by the Company so far as appears
from our examination of those books, and proper returns
adequate for the purposes of our audit have been received
from the Branches of the Company. In our opinion, and to
the best of our information and according to the explanations
given to us the Balance Sheet exhibits a true and fair view
of the state of the Company's affairs as at 31st December,
1951, and the Profit and Loss Account shows a true and fair
view of the profits for the year ended on that date.

Buckley Hughes
Auditor.

Maurice McNamara.
Managing Director.

6th March, 1952.

Totals

Shareholders Funds   62,022 5 7
Current Liabilities   381,534 14 3

                                  £443,556 19 10

& CO. PTY. LTD.
AS AT 31ST DECEMBER, 1951.

	£. s. d.	£. s. d.

ASSETS

Fixed Assets

Properties, Improvements, Plant, etc.

	£ s d	£ s d
Office Premises		
Additions	21,700 0 0	
Depreciation	836 0 0	20,864 0 0
Saleyards		
Value at 31.12.50	5,977 0 9	
Additions	593 9 0	
	6,570 9 9	
Sales	28 0 0	
	6,542 9 9	
Depreciation	2,882 4 6	3,660 5 3
Staff Residences		
Value at 31.12.50	3,579 8 8	
Additions	13,959 17 9	
	17,539 6 5	
Depreciation	1,116 13 4	16,422 13 1
Office Furniture and Fittings		
Value at 31.12.50	3,185 12 3	
Additions	2,268 6 8	
	5,453 18 11	
Depreciation	416 19 2	5,036 19 9
Office Machines		
Additions	24 11 0	
Depreciation	1 4 6	23 6 6
Motor Vehicles		
Value at 31.12.50	11,562 14 3	
Additions	16,575 4 11	
	28,137 19 2	
Sales	4,596 5 0	
	23,541 14 2	
Depreciation	6,423 15 5	17,117 18 9
Total Fixed Assets		63,125 3 4
Intangible Asset		
Goodwill		
Value at 31.12.50	4,000 0 0	
Additions	35,000 0 0	
	39,000 0 0	
Less Written Off	4,000 0 0	35,000 0 0
Carried Forward —		98,125 3 4